From our home on planet Earth, we can look up at the night sky and see many stars. But there is so much more to space than stars. Read on to learn about some of the amazing objects that make up our universe.

Asteroid

An asteroid is a small, rocky object that orbits our sun. Asteroids are smaller than planets, ranging in size from meters to hundreds of kilometers. Most asteroids are found in the asteroid belt, an area of space between Mars and Jupiter.

Black Hole

A black hole is formed when a very large star collapses into itself at the end of its life cycle. All of the star's matter is then squeezed into a tiny space, which produces a huge amount of gravity. The gravity is so strong that nothing can escape it, not even light.

Comet

A comet is like a dirty snowball that orbits the sun. It is made of rock, dust, and frozen gasses that get heated up when it gets close to the sun. The heating of the gasses is what produces the tail comets are famous for.

Double Star

A double star, also known as a binary star, occurs when two stars orbit around a fixed point between them. They are bound together by gravity. They usually look like one star, but can be revealed to be two stars with powerful telescopes.

Exoplanet

An exoplanet is a planet that orbits around a star outside of our solar system. Since this planet is found outside of our solar system, it is called an extrasolar planet. As of 2020, over 4,000 exoplanets have been discovered.

Flare Star

A flare star is a star that varies in brightness for a few minutes. The change in brightness is due to outbursts of radiation. The change in brightness is random and sudden.

Galaxy

A galaxy is a system composed of billions of stars bound together by gravity. Galaxies come in different shapes and sizes, including spiral, elliptical, and irregular. Our solar system is located in the Milky Way Galaxy.

Heliosphere

The heliosphere is the part of space that contains the sun and all planets, and extends to the boundary of where the solar wind reaches. The heliosphere ends at the heliopause, which is where the interstellar medium begins.

Interstellar Medium

The interstellar medium is the gas, dust, and radiation that exists between star systems. It consists mostly of hydrogen gas, with helium and trace amounts of other gases. The space it occupies is known as interstellar space.

Jet

A jet is a stream of radiation and particles that are ejected from black holes or stars. Most matter is attracted by a black hole's strong gravity. A jet, however, consists of some particles shot out in a narrow beam at almost the speed of light.

Kuiper Belt

The Kuiper Belt is a disk-shaped region of space beyond Neptune that consists of many small, icy objects. The objects are believed to be left over from when the solar system formed. Three dwarf planets, Pluto, Haumea, and Makemake, are found in the Kuiper Belt.

Local Group

The Local Group is a group of galaxies that our galaxy, the Milky Way, is found in. The Milky Way and the Andromeda Galaxy are the two largest galaxies of the Local Group. There are over 50 galaxies that make up the Local Group.

Meteoroid

A meteoroid is a rocky object smaller than an asteroid that orbits the sun. If a meteoroid enters and burns up in Earth's atmosphere, it is called a meteor. If it enters Earth's atmosphere and instead of burning up, it reaches Earth's surface, it is called a meteorite.

Nebula

A Nebula is an interstellar cloud of dust and gas. It contains the ingredients for a star to start its life cycle. The gas and dust eventually clump together to form denser and denser regions to become stars, and after that, planets.

Oort Cloud

The Oort Cloud is a spherical region of space beyond the Kuiper Belt that surrounds our solar system. It consists of many icy objects. It is also believed to be where comets come from.

Pulsar

A pulsar is a spinning neutron star that emits electromagnetic radiation beams from its poles. A neutron star is what is left after a very large star explodes. It is made up entirely of a mass of neutrons confined to a very small space, creating a strong gravitational pull.

Quasar

A quasar is an extremely large and bright object found in the remote areas of space. They are powered by super massive black holes, emit enormous amounts of energy, and are some of the oldest objects in the universe. They occur in the center of galaxies and contain jets.

Red Giant and Supergiant

Red Giants and Red Supergiants occur when a star has no more hydrogen left in its core to fuse into helium. The star then expands to fifty times its normal size and cools in temperature. Average size stars become Red Giants and massive stars become Red Supergiants.

Supernova

A supernova is a powerful and bright explosion of a Red Supergiant that occurs when the star can no longer sustain nuclear fusion. This explosion ejects most of the star's mass. This explosion also creates the heavy elements of the universe.

Trans-Neptunian Object

A trans-Neptunian object is a dwarf planet that orbits the sun and lies beyond Neptune. They are made of rock and ice. Examples include Eris, Pluto, Haumea, and Makemake.

Universe

The universe is everything that exists in space and time. The universe is estimated to be 13.8 billion years old. In addition to containing all the matter we know exists, the universe also contains more mysterious elements known as dark matter and dark energy.

Variable Star

A variable star is a star whose brightness changes over time. The change in brightness can sometimes be traced to an object that moves in front of the star from our vantage point on Earth. Instead of changing brightness, some variable stars change color over time.

White Dwarf

A white dwarf is the last stage of an average size star's life cycle. Once the star has used up all of its fuel, it collapses in on itself and forms a planetary nebula. What is left after that in the core is a small, white star.

X-ray Star

An X-ray binary star is a bright two-star system that gives off X-ray radiation. The X-rays are produced when matter moves from the lower mass star to the higher mass star. They appear very bright in X-rays.

Yellow Dwarf

A yellow dwarf is a star average in size and temperature, much like our sun. It is also known as a G-type main sequence star. It is in the middle of its life cycle, and is therefore stable.

Zero Age Main Sequence Star

A zero age main sequence star is when a star is born. This occurs when a stellar nebula becomes dense enough at the core to start burning and fusing hydrogen, creating a star.

www.ingramcontent.com/pod-product-compliance
Lightning Source LLC
Chambersburg PA
CBHW051943210526

45473CB00006B/2365